THE INNER WORKINGS OF A WATCH

A SIMPLE GUIDE FOR ENTHUSIASTS OF CLOCKWORK MECHANISMS

British Library Cataloguing-in-Publication Data
A catalogue record for this book is available from the
British Library

Contents

A History of Clocks and Watches

Horology (from the Latin, Horologium) is the science of measuring time. Clocks, watches, clockwork, sundials, clepsydras, timers, time recorders, marine chronometers and atomic clocks are all examples of instruments used to measure time. In current usage, horology refers mainly to the study of mechanical time-keeping devices, whilst chronometry more broadly included electronic devices that have largely supplanted mechanical clocks for accuracy and precision in time-keeping. Horology itself has an incredibly long history and there are many museums and several specialised libraries devoted to the subject. Perhaps the most famous is the Royal Greenwich Observatory, also the source of the Prime Meridian (longitude 0° 0' 0"), and the home of the first marine timekeepers accurate enough to determine longitude.

The word 'clock' is derived from the Celtic words clagan and clocca meaning 'bell'. A silent instrument missing such a mechanism has traditionally been known as a timepiece, although today the words have become interchangeable. The clock is one of the oldest human interventions, meeting the need to consistently measure intervals of time shorter than the natural units: the day, the lunar month and the year. The current sexagesimal system of time measurement dates to approximately 2000 BC in Sumer. The Ancient Egyptians divided the day into two twelve-hour periods and used large obelisks to track the movement of the sun. They also developed water clocks, which had also been employed frequently by the Ancient Greeks, who called them 'clepsydrae'. The Shang Dynasty is also believed to have used the outflow water clock around the same time.

The first mechanical clocks, employing the verge escapement mechanism (the mechanism that controls the rate of a clock by advancing the gear train at regular intervals or 'ticks') with a foliot or balance wheel timekeeper (a weighted wheel that rotates back and forth, being returned toward its centre position by a spiral), were invented in Europe at around the start of the fourteenth century. They became the standard timekeeping device until the pendulum clock was invented in 1656. This remained the most accurate timekeeper until the 1930s, when quartz oscillators (where the mechanical resonance of a vibrating crystal is used to create an electrical signal with a very precise frequency) were invented, followed by atomic clocks after World War Two. Although initially limited to laboratories, the development of microelectronics in the 1960s made quartz clocks both compact and cheap to produce, and by the 1980s they became the world's dominant timekeeping technology in both clocks and wristwatches.

The concept of the wristwatch goes back to the production of the very earliest watches in the sixteenth century. Elizabeth I of England received a wristwatch from Robert Dudley in 1571, described as an arm watch. From the beginning, they were almost exclusively worn by women, while men used pocket-watches up until the early twentieth century. This was not just a matter of fashion or prejudice; watches of the time were notoriously prone to fouling from exposure to the elements, and could only reliably be kept safe from harm if carried securely in the pocket. Wristwatches were first worn by military men towards the end of the nineteenth century, when the importance of synchronizing manoeuvres during war without potentially revealing the plan to the enemy through signalling was increasingly recognized. It was clear that using pocket watches while in the heat of battle or while mounted on a horse was impractical, so officers began to strap the watches to their wrist.

The company H. Williamson Ltd., based in Coventry, England, was one of the first to capitalize on this opportunity. During the company's 1916 AGM it was noted that '...the public is buying the practical things of life. Nobody can truthfully contend that the watch is a luxury. It is said that one soldier in every four wears a wristlet watch, and the other three mean to get one as soon as they can.' By the end of the War, almost all enlisted men wore a wristwatch, and after they were demobilized, the fashion soon caught on - the British *Horological Journal* wrote in 1917 that '...the wristlet watch was little used by the sterner sex before the war, but now is seen on the wrist of nearly every man in uniform and of many men in civilian attire.' Within a decade, sales of wristwatches had outstripped those of pocket watches.

Now that clocks and watches had become 'common objects' there was a massively increased demand on clockmakers for maintenance and repair. Julien Le Roy, a clockmaker of Versailles, invented a face that could be opened to view the inside clockwork – a development which many subsequent artisans copied. He also invented special repeating mechanisms to improve the precision of clocks and supervised over 3,500 watches. The more complicated the device however, the more often it needed repairing. Today, since almost all clocks are now factory-made, most modern clockmakers *only* repair clocks. They are frequently employed by jewellers, antique shops or places devoted strictly to repairing clocks and watches.

The clockmakers of the present must be able to read blueprints and instructions for numerous types of clocks and time pieces that vary from antique clocks to modern time pieces in order to fix and make clocks or watches. The trade requires fine motor coordination as clockmakers must frequently work on devices with small gears and fine machinery, as well as an appreciation for the original art form. As is evident from this very short history of clocks and watches, over the centuries the items themselves have changed – almost out of recognition, but the importance of time-keeping has not. It is an area which provides a constant source of fascination and scientific discovery, still very much evolving today. We hope the reader enjoys this book.

THE PARTS OF A WATCH

With high-class watches, one can expect a very close rate of time under both extreme and normal conditions, but the inexpensive watch, by reason of its condition, cannot be expected to keep time within several seconds a day. A keen student, however, will soon be able to classify the various grades and execute work accordingly. Any new parts should be faithfully copied, if it is not possible to obtain standard material, in order to maintain the standard of the watch. Good work always reflects credit on the repairer.

FIG. 43.—A MODERN SWISS LEVER MOVEMENT OF THE POPULAR 10 1/2 LIGNE SIZE (SHOWN GREATLY ENLARGED).

Special Names for Parts.—The first step is to become thoroughly acquainted with the numerous components of an ordinary watch. Many parts have special names, and to be conversant with them will often save considerable time when ordering new material. The enlarged illustration on the opposite page depicts a modern Swiss Lever movement of the popular 10 1/2-ligne size.

"Lignes" and "sizes" are the measurements usually used to determine the size of a movement. In Fig. 44 are shown the various diameters of a movement. Of the two main dimensions that of the largest diameter is usually taken, and the most common measurement is the ligne. As 1 ligne equals approximately 3/32 in. a 10 1/2-ligne watch measures 15/16 in. which is short of an inch. The American industry favours the "size" as a unit of measurement. Size O equals 1 5/30 in. Each size above size O increases by 1/30 in., and a size below O decreases by 1/30 in. 10 1/2-ligne movements are to be found in both gentlemen's and ladies' wrist-watches. Until a few years ago this size was almost universal in ladies' watches, and the cheaper watches still favour this size.

FIG. 44.—THE VARIOUS DIAMETERS OF A MOVE-
MENT.

FIG. 45.—THREE DIFFERENT TYPES OF SCREW USED IN THE CONSTRUCTION OF A WATCH.

Number of Screws.—Some watches have as many as 150 separate pieces, and of this large number there are at least 35 screws. Fig. 45 shows three different types of screw: the cheese-headed plate screw, the flat-headed case screw, and the small jewel screw with countersunk head. The main frame of the movement consists of two plates: the bottom or dial plate and the top or back plate, which is visible when the case is opened. The modern back plate has changed from a circular plate into a number of sections usually called bars or bridges, thereby rendering the works easily accessible.

Almost half the movement is used for the large bar that supports the mainspring barrel, a thin cylindrical metal box with teeth around the outside edge. The barrel is fitted with a cover, and the axle upon which it rotates is called the arbor. The arbor has a short hook which engages the inner eye of the mainspring.

The Great Wheel.—The barrel, or main driving wheel, is often referred to as the great wheel, and the other wheels are arranged in the following order. In the centre of the movement and driven by the barrel is the centre wheel; next, the 3rd wheel; then the 4th wheel (the seconds wheel); and finally the 5th wheel (the escape wheel). The remaining section is known as the escapement. When referring to the escapement, this is generally assumed to include the escape wheel, the small anchor-shaped piece called the lever, which arrests and releases the escape wheel tooth by tooth, and the balance and its kindred pieces.

The balance wheel is mounted on a slender axle—the balance staff. Fixed upon the staff above the balance wheel is the hair-spring and below the balance wheel is the roller. The roller is fitted with a small impulse pin, but when a jewelled pin is used it is commonly called the ruby pin. The function of the roller is to unlock the pallets. Fig. 46 shows the balance, balance staff and roller, and the position of the fork of the pallets with regard to the ruby pin.

Jewels.—In jewelled watches, the most popular number of jewels is 15. These jewels are not mere ornaments, but are used to minimise wear. The 15 jewels are always arranged in this order. Two each for the 3rd, 4th, and 5th wheels and pallets, 4 for the balance, 2 pallet stones and the ruby pin. Fig. 49 shows sections of plate and balance jewels. No. 1 is a section of the jewels used for 3rd, 4th, and escape wheels;

No. 2 is an endstone; and No. 3 shows the arrangement of the balance jewels, one at each end of the balance staff. In high-class watches, jewel hole and endstone are fixed in separate settings and kept in position by 2 jewel screws as shown. It will be observed that the balance jewel hole differs slightly from the ordinary jewel hole. For example, the oil sink is inside on the balance hole and outside on the plate hole.

The pivots, the short projections of the pinions and staffs which actually rotate in the bearings also differ in shape. In Fig. 48 are depicted at *A* an ordinary pivot with a square shoulder and at *B* a balance pivot with a conical shoulder. Type *A* pivots are used with No. 1 type jewels. In high-class watches the lever and escape wheel pivots are often made conical and provided with balance type jewels and endstones.

The Bottom Plate.—Fig. 50 shows a bottom plate. This carries the small winding and hand-setting wheels and the levers that operate them. The winding shaft passes through two small wheels seen at the top, the top or crown wheel engages the smaller of the 2 flat steel wheels seen in Fig. 43 at right-angles. When winding, the mainspring is prevented from "running back" by the action of the pawl or click which arrests the larger of the winding wheels.

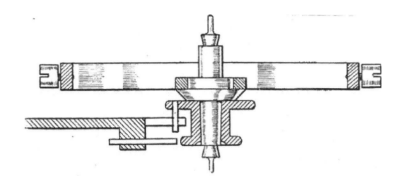

FIG. 46.—THE BALANCE, BALANCE STAFF AND
ROLLER, AND THE POSITION OF THE FORK OF
THE LEVER WITH REGARD FO THE RUBY PIN.

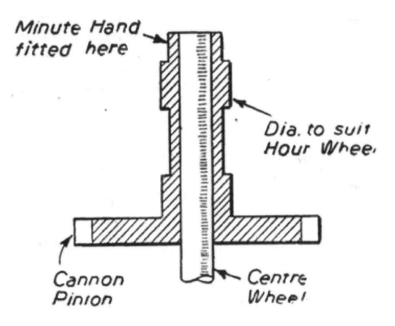

FIG. 47.—A CANNON PINION.

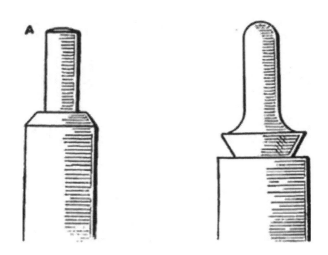

FIG. 48.—A, AN ORDINARY PIVOT, WITH A SQUARE
SHOULDER; AND B, A BALANCE PIVOT WITH A
CONICAL UNDERCUT SHOULDER.

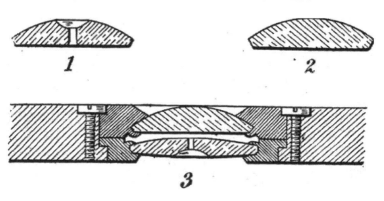

FIG. 49.—SECTIONS OF PLATE AND BALANCE JEW-
ELS.

14

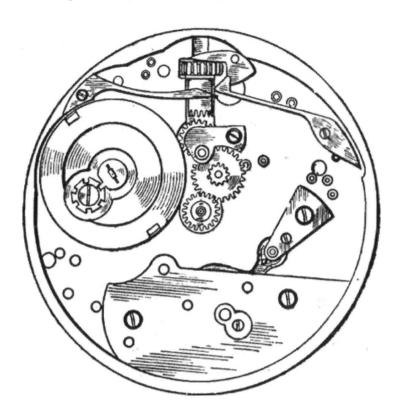

FIG. 50.—A BOTTOM PLATE SHOWING THE PARTS BEHIND THE DIAL. THIS MOVEMENT IS A PATEK PHILLIPPE.

The winding shaft is prevented from being pulled right out by the pull-up piece, which serves the dual purpose of retaining the winder and forcing down, by means of the return lever, the lower wheel on the winding shaft, causing it to engage the intermediate hand setting wheel. The intermediate wheel gears with the minute wheel—a flat brass wheel having a short pinion and rotating on a stud fixed in the plate—and the teeth of the minute wheel gear with those of the cannon pinion. The cannon pinion is really a small tube with teeth around the bottom and it fits friction tight on the extended pivot of the centre wheel. It is upon this tubular pinion that the minute hand is fixed.

When the winder is pulled out, it depresses the lever and the castle wheel, and the motion wheels (the hand wheels), are engaged and can be turned around to the desired position, Fig. 47 shows the cannon pinion.

THE COMPENSATING BALANCE &
HAIRSPRING

To illustrate the general principles of timing, it may be of interest first to make a comparison between the balance in a watch and the pendulum in a clock, as both of them evidently perform the function of measuring or beating time. The pendulum, as we all know, requires no special spring to bring it to its centre line, the perpendicular, as the force of gravity furnishes the necessary power for doing this work in a very ideal way. When a pendulum is put in motion, it makes a vibration in a certain interval of time, and in proportion to its length, regardless of its weight, because the force of gravity acts on it in proportion to its mass. The length of a pendulum is reckoned from its centre of suspension to its centre of oscillation, which latter point is located a short distance below the middle of the bob. If a weight is added *above this point*, the clock will *gain*, because it raises the centre of oscillation and has the same effect on the time-keeping as raising the whole bob, which is equivalent to a shortening of the pendulum; but if a weight is added *below this point, it has the opposite effect*, as it really lengthens the pendulum. Reasoning from these facts we come to the conclusion that we can make a certain change in the rate of a clock in three different ways. For example, we may make it gain: (1) by raising the bob; (2)

by adding weight above the centre of oscillation; and (3) by reducing the weight below that point. An interesting fact in relation to the pendulum, which may not be generally known among watchmakers, is that its rate of vibration varies slightly with change of latitude, and also of altitude (that is, its height above the sea-level), making a clock lose at the Equator and at high altitudes, and gain as we go nearer the sea-level and the Poles. This is due, partly to the distance from the centre of the earth, which is greater at the Equator than at the Poles, and partly to the centrifugal force resulting from the rotation of the earth on its axis. Both these factors tend to make an object weigh less (on a spring balance) at the Equator than at the Poles, and also cause a change in the rate of a clock as stated above. In view of these facts we might, as a fourth way of making a clock gain—although not a very practical one— move to a locality nearer the Pole. A balance is different from a pendulum in three fundamental points: first, it is poised; consequently the force of gravity has no effect on it, except as its influences the friction on its pivots; second, the vibrations are controlled by a spring instead of the force of gravity; third, a weight (mass) added to a balance will always make it vibrate slower, provided it is not thereby put out of poise, and the retarding effect will be greater the farther the weight is placed away from its centre. One difficulty encountered in the first attempt to make accurate timepieces was the variation in the dimensions of metals caused by difference in temperature.

All metals with the exception of a recently discovered alloy of steel and nickel (64 parts of steel and 36 of nickel) have the property of expanding with increase of temperature—the different metals showing a somewhat different rate of change. As the length of the pendulum is the all-important factor in the timing of clocks, so also is the diameter of the balance and the length and resiliency of the hairspring in a watch. It is absolutely necessary to devise some means of compensating for changes in temperature before a reliable timepiece of either form can be made. So far as this problem applies to clocks, the mercury pendulum proves to be a very satisfactory solution, at least so far as accuracy is concerned. The bob of this pendulum is composed of one or more tubes of glass or iron, and these tubes are filled with mercury to a certain height. When of proper dimensions, the expansion and contraction of this column of mercury raises or lowers its mass to exactly compensate for the change in the length of the pendulum rod due to variations in the temperature. This method, although very satisfactory for clocks, cannot, of course, be applied to watches, for obvious reasons, but for this purpose we make use of the property of the metals alluded to above, namely, *the difference* in the ratio of *expansion in different metals.*

STEEL

BRASS

FIG. 51.—TWO BARS OF EQUAL LENGTH WHEN AT NORMAL TEMPERATURE. THE DOTTED LINES IN-DICATE THE RELATIVE EXPANSIONS WHEN EACH IS HEATED TO A SIMILAR DEGREE.

STEEL
BRASS

FIG. 52.—THE EFFECT OF HEAT ON THE BI-METAL-LIC BAR. THE GREATER EXPANSION OF THE BRASS CAUSES THE BAR TO CURVE UPWARD.

STEEL
BRASS

FIG. 53.—THE TWO METALS FUSED TOGETHER.

STEEL
BRASS

FIG. 54.—IF COLD WERE APPLIED INSTEAD OF HEAT, THE BAR WOULD CURVE IN THE OPPOSITE DIRECTION.

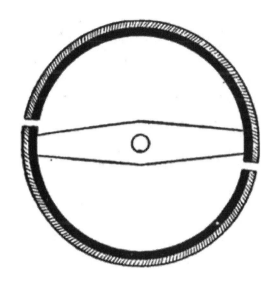

FIG. 55.—WATCH BALANCE IN WHICH THE ARMS AND INNER PORTION OF THE RIM ARE OF STEEL. THE OUTER PORTION IS BRASS FUSED TO THE STEEL. THE RIM IS SEVERED AT TWO POINTS NEAR THE ARM, PERMITTING THE RIM TO MOVE UNDER CHANGE OF TEMPERATURE.

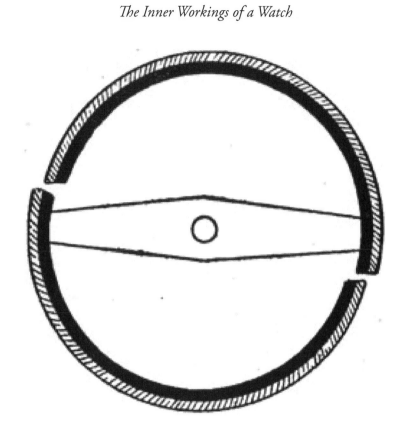

FIG. 56.—THE BALANCE UNDER THE INFLUENCE OF HEAT. IT WILL BE NOTICED THAT THE FREE ENDS OF THE BALANCE RIM HAVE CURVED IN-WARDS, THUS REDUCING THE DIAMETER OF THE BALANCE (RADIUS OF GYRATION).

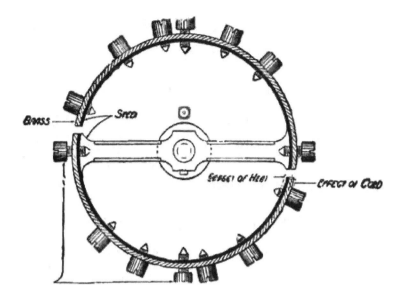

FIG. 57.—DIAGRAM SHOWING THE BALANCE,
LOADED WITH SCREWS FOR TIMING AND ADJUST-
MENT AND COMPENSATION.

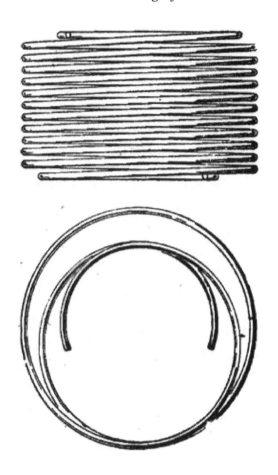

FIG. 58.—HELICAL BALANCE SPRING FITTED USU-ALLY TO MARINE CHRONOMETERS.

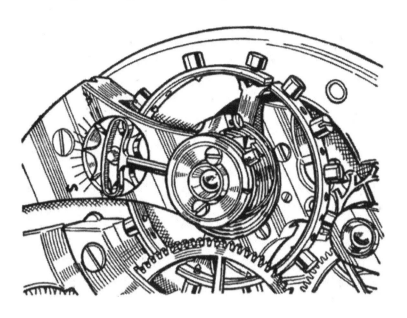

FIG. 59.—THE WALTHAM MICROMETRIC REGU-
LATOR. OTHER MAKERS USE AN EXTERNAL RACK
AND PINION, OR A SPRING-LOADED CAM.

FIG. 60.—AN UP-TO-DATE BIMETALLIC MARINE CHRONOMETER BALANCE, MADE FROM BRASS AND STEEL.

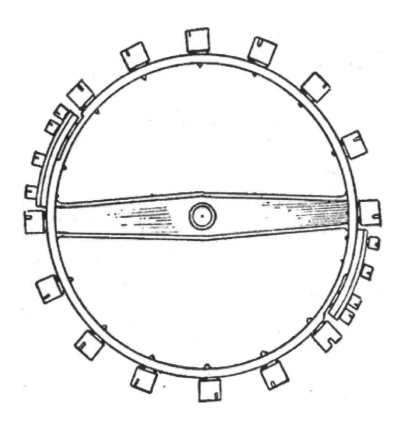

FIG. 61.—THE "AFFIX" DITISHEIM MONO-ME-
TALLIC UNCUT BALANCE. COMPENSATION IS
SECURED BY THE TWO SMALL AUXILIARY PIECES,
AND THE USE OF AN ELINVAR HAIRSPRING.

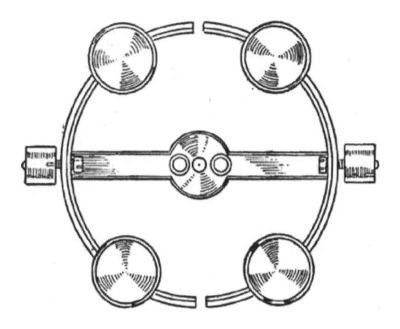

*FIG. 62.—THE GUILLAUME BALANCE FOR CHRO-
NOMETERS. IT IS, OF COURSE, BI-METALLIC AND
CUT. MADE FROM BRASS AND NICKEL STEEL, IT
ALMOST GETS RID OF MIDDLE TEMPERATURE ER-
ROR.*

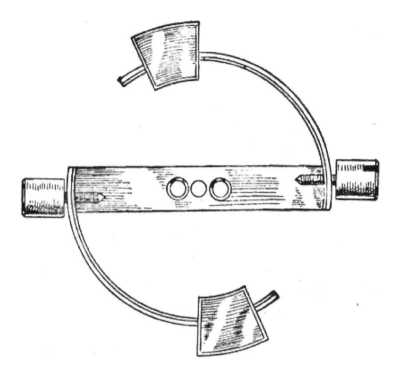

FIG. 63.—MARINE CHRONOMETER BI-METALLIC
BALANCE AS INTRODUCED BY EARNSHAW.

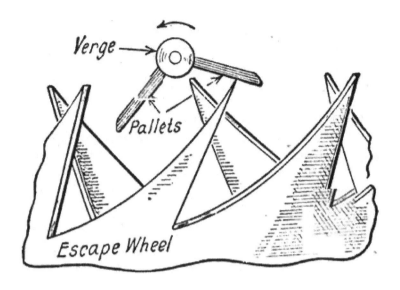

FIG. 64.—THE VERGE ESCAPEMENT—NOW OBSO-LETE.

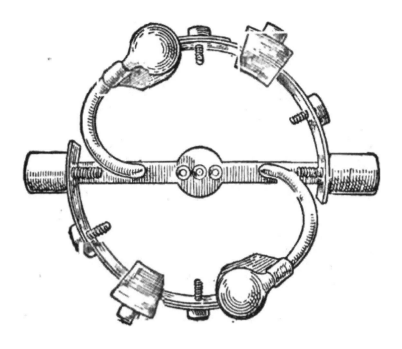

FIG. 65.—THE LOSEBY COMPENSATING BALANCE;
ON EACH ARM OF THE RIM IS A VESSEL CONTAIN-
ING MERCURY.

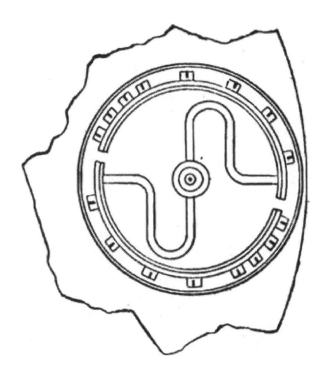

*FIG. 66.—THE WYLER COMPENSATING BALANCE
AS FITTED TO THE WYLER SELF-WINDING WRIST-
WATCH. IT IS SURROUNDED BY A PROTECTING
RIM, AND IS SHOCK-PROOF.*

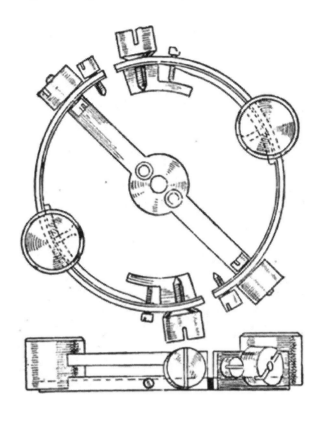

*FIG. 67.—THE AUXILIARY COMPENSATION BAL-
ANCE INVENTED BY KULLBERG. THE RIM IS OF
THE USUAL BI-METALLIC TYPE. THE STEEL FOR
80° ON EACH SIDE OF THE BALANCE IS MADE
THICKER. ITS OBJECT IS TO OVERCOME MIDDLE
TEMPERATURE ERROR. USUALLY FITTED TO MA-
RINE CHRONOMETERS.*

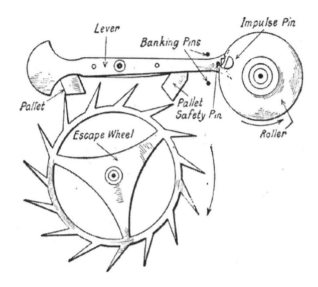

FIG. 68.—THE ENGLISH OR SIDE LEVER. THIS FORM OF ESCAPEMENT IS RAPIDLY BEING SUPERSEDED BY THE "STRAIGHT LINE" ESCAPEMENT, IN WHICH THE CENTRES OF THE BALANCE, LEVER AND ESCAPE WHEEL ARE IN LINE. THE SPUR TOOTH ESCAPE WHEEL HAS ALMOST ENTIRELY BEEN REPLACED BY THE CLUB TOOTH ESCAPE WHEEL MADE OF STEEL, NOT BRASS AS IS THE SPUR WHEEL.

A number of holes are drilled radially through the bimetallic rim, and these holes are tapped to receive the balance screws. Usually about twice as many holes are made in the rim as the number of screws used in the balance; this is done to give opportunity for moving the scfews in the final adjusting to temperatures. The object in using screws in the balance rim is two-fold; first, to provide the necessary weight (mass) in the rim, and second, to have this weight movable for temperature adjustments, as stated above.

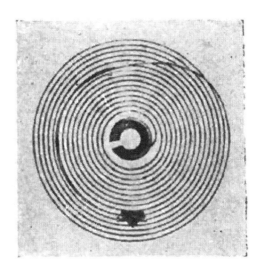

FIG. 69.—THE OVERCOIL HAIRSPRING, THE OUTER CURVE OF WHICH IS MADE TO CONFORM WITH PHILLIP'S THEORY. IT GIVES A CONCENTRIC AC-TION TO THE HAIRSPRING AND THUS REDUCES POSITION ERRORS. SUCH AN OVERCOIL IS OFTEN ERRONEOUSLY CALLED "BREGUET," WHICH IS A SPECIALLY FORMED CURVE. IT IS IMPOSSIBLE TO ELIMINATE POSITION ERRORS WITH A FLAT HAIR-SPRING.

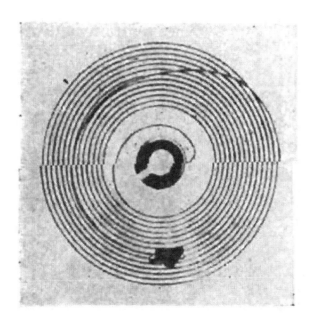

FIG. 70.—THE WALTHAM HAIRSPRING SHOWING INNER AND OUTER TERMINAL CURVES. THE INNER CURVE SHOWN IN FIG. 69 TENDS TO THROW THE SPRING OUT OF POISE DURING ITS VIBRATIONS. WITH THE INNER CURVE SHOWN ABOVE, THIS TENDENCY IS ELIMINATED. WALTHAM HAIRSPRINGS ARE HARDENED AND TEMPERED IN FORM. MOST OTHER SPRINGS ARE FORMED AFTER THE SPRINGS HAVE BEEN TEMPERED.

We will now understand from what has been said that when a compensating balance is exposed to a higher temperature, every part of it expands, or grows larger, but as a result of the combination of the two metals in the rim, and the ends of the rim being free to move, each half of the rim will curve inward, carrying its weight towards the centre of the balance, and thus compensate for the lengthening of the arms and the weakening of the hairspring. If a balance is exposed to a lower temperature, the action will, of course, be in the opposite direction.

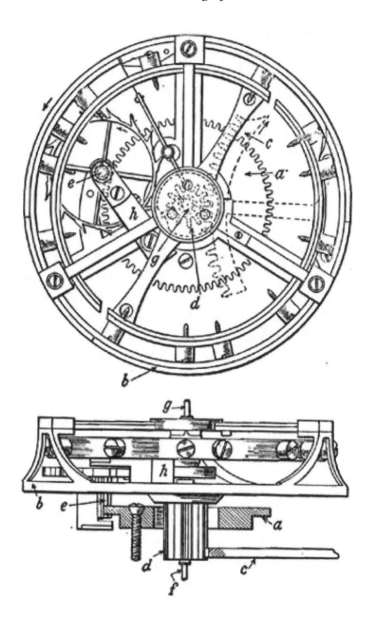

*FIG. 71.—PLAN AND ELEVATION OF THE TOURBIL-
LON ESCAPEMENT. THE BALANCE IS MOUNTED
ON A CARRIAGE WHICH ITSELF REVOLVES ONCE
EVERY MINUTE, THUS ELIMINATING POSITION
ERRORS. IN THIS DIAGRAM A IS THE FOURTH
WHEEL, B THE REVOLVING CAGE OR CARRIAGE, D
CARRIAGE PINION, E THE ESCAPE PINION, F PIVOT
FOR SECONDS HAND, H ESCAPE COCK, G UPPER
PIVOT, C THE THIRD WHEEL GEARING WITH D.
THE KARRUSEL, INVENTED BY BONNIKSEN, IS
SOMEWHAT SIMILAR EXCEPT THAT IT ROTATES
ONCE IN 52 1/2 MINUTES.*

When a watch is to be adjusted to temperatures, it is run 24 hours, dial up, in a temperature of 90°F., and its rate compared with a standard. It is then run 24 hours, dial up, in a temperature of 40°F. If its shows a gain in the 40° temperature, as compared with the running in the 90°, it is said to be under - compensated. This is remedied by moving some screws nearer the free ends of the rim. This will, of course, result in a greater compensating effect, because the screws which we move nearer the ends of the rim must travel a greater distance in or out in relation to the centre of the balance when the balance is exposed to changes of temperature. After the screws have been moved, the movement is tried again the same length of time, and so on, until it runs the same in both temperatures. When a screw is moved in one side of the balance, it is, of course, necessary that the corresponding screw in the other side should be moved the same. A modern compensation balance, combined with a correctly proportioned steel Breguet hairspring, which has been hardened and tempered in form, constitute a time-measuring device of marvellous accuracy. And the bi-metallic rim, hardened as the Waltham balances are, so as to be perfectly safe against distortion from ordinary handling, is certainly a boon to the watchmaker.

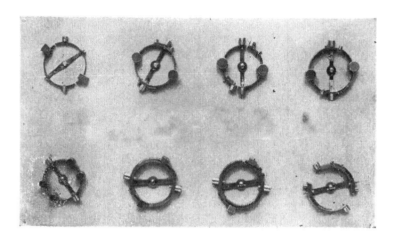

FIG. 73A.—EIGHT EXAMPLES OF COMPENSATING BALANCES FOR MARINE CHRONOMETERS.

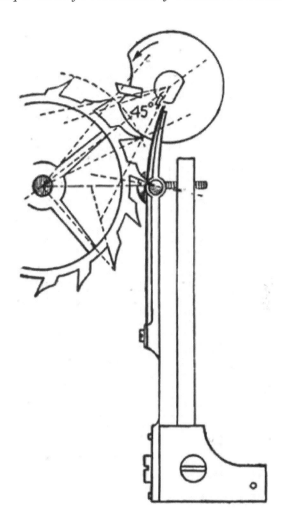

FIG. 72.—ANOTHER TYPE OF MARINE CHRONOM-
ETER ESCAPEMENT, AS DESIGNED BY EARNSHAW.
THE DISADVANTAGE IS THAT IT TENDS TO "SET"
OR STOP WHEN WORN IN THE POCKET.

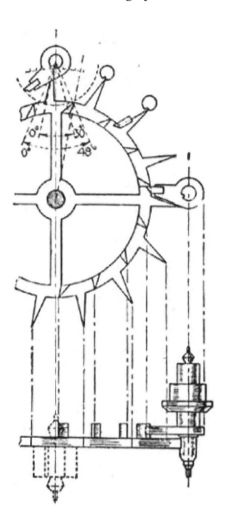

FIG. 73.—THE DUPLEX ESCAPEMENT. THIS TYPE WAS EXTREMELY POPULAR AT ONE TIME, AND WAS FINALLY FITTED TO THE CHEAP WATERBURY WATCHES. IT IS NOW OBSOLETE.

The mean time screws used in the Waltham balances furnish an excellent means for accurate timing, as two, on opposite sides, can be turned an equal amount in (making the watch run faster) or out (slower) without changing the poise of the balance.

The following is the approximate effect of *one-half turn* of two mean time screws:

18-size and 16-size	..	..	$2\frac{1}{4}$ seconds per hour
Colonial series and 12-size		..	$2\frac{1}{2}$,, ,,
0 size and 3-0-size	..	..	3 ,, ,,
Jewel series	..	..	$3\frac{1}{4}$,, ,,
10 Ligne	..	..	$3\frac{1}{2}$,, ,,

An illustration of a hairspring supplied with a theoretically correct outer terminal, commonly known to watchmakers by the name of Overcoil, is shown in Fig. 69.

As is well known to watchmakers, hairsprings are supplied with overcoils to secure a concentric action of the hairspring while the balance is in motion. A concentric action of the hairspring is necessary, in order to reduce the position error.

This result is partially obtained when a hairspring is supplied with a theoretically correct outer terminal or overcoil, whereby the centre of gravity of the outer half of the hairspring is made to coincide with the centre of the balance at every stage of its vibration.

In a watch fitted with a hairspring with an outer terminal curve only, there still remains a position error, part of which, at least, is due to the fastening of the inner end of the spring, which, in its ordinary form, tends to throw the spring out of poise during its vibrations.

This the Waltham Watch Company has succeeded in overcoming by making hairsprings with theoretically correct inner terminal curves. This inner curve maintains the body of the spring in perfect poise with the balance, during both its opening and closing vibrations. I show an illustration (Fig. 70) of a hairspring of this kind with inner and outer terminals.

The design of the hairspring having been perfected, there came the problem of properly producing these springs.

Flat hairsprings, that is to say, those without an overcoil, are only fitted to cheaper watches, although high-grade thin dress watches are sometimes fitted with them in order to reduce the thickness of the movements. It is impossible to eliminate position errors when a flat hairspring is fitted.

THE LEVER ESCAPEMENT

The proper action of the human heart is no more essential to ensure a sound and healthy body, than is the proper adjustment and action of the escapement to the reliable performance of the watch. If the watch escapement is properly made and adjusted it will not only run—but will run with marvellous accuracy. So the timekeeping qualities of the watch are in large measure dependent on the condition of the escapement. It is therefore of great importance that every watchmaker should acquire an intimate knowledge of all the actions that are involved in the kinds of escapements with which he has anything to do.

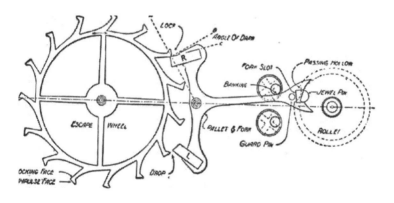

FIG. 74.—ALL GOOD WATCHES HAVE A LEVER ES-CAPEMENT OF THE TYPE SHOWN HERE.
NOTE THE JEWELLED PALLETS.

The only kind of escapement used in modern watches is the detached lever escapement, sometimes designated as the anchor escapement. This escapement requires no special introduction to watchmakers, for by extensive use, and by the test of time, it has been proved to be the most practical, as well as the most reliable form of escapement for pocket timepieces.

I include in this chapter some drawings of the lever escapement that they may be convenient for reference, and an aid to the clear understanding of the text.

The function of the escapement is to impart to the balance, regularly, and with as small loss as possible, the power which has been transmitted through the train from the mainspring to the escape pinion. In the lever escapement this is accomplished by means of two distinct actions: first, the action of the escape wheel and pallet; second, the action of the fork and roller pin. Fig. 74, is a plan view of the Waltham Lever Escapement, as used in the 16-size, and the 18-size models; movements drawn to scale 5-to-1, and giving the names of the principal parts and features of the same. The escape wheel is mounted friction tight on the slightly tapered staff of the escape pinion. It has 15 teeth, called "Club Teeth" on account of their peculiar shape, resulting from the addition of impulse faces to the ends of the teeth, and to distinguish them from "ratchet teeth", the name given to a style of pointed teeth used on escape wheels in an earlier form of lever escapement. In descriptions of this escapement the term "exposed pallets" is used. This means that the pallet stones are visible, with the active ends standing out free from the body of the pallet, as distinguished from an earlier form of pallet with "covered stones" set in slots running in the plane of the pallet.

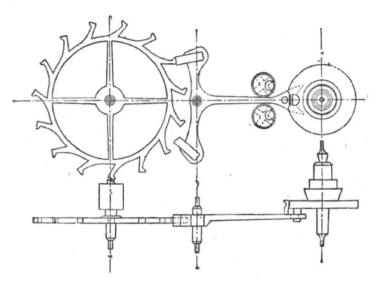

FIG. 75.—*LEVER ESCAPEMENT OF THE SINGLE ROLLER TYPE.*

The Pallet Action.—The action of the escape wheel and pallet includes the following features: impulse, drop, lock, draft and slide, and in giving a general description of these actions we will consider briefly what constitutes each one of these features. The pallet is of the kind called "circular pallet", which means that the distance from the pallet arbor to the middle of the impulse face is the same for both pallet stones. Another kind of pallet is made with "equidistant lock", that is, the distance from the pallet arbor to the point where the lock takes place, is the same for both pallet stones. The pallet is mounted on its arbor, which is located close to the periphery of the escape wheel. A theoretically correct distance in relation to the diameter of the escape wheel will not allow an excess of clearance between the pallet and the escape wheel teeth when opposite the pallet arbor, and for that reason the amount of stock in the pallet is made very small at that point. The pallet is slotted for the 2 pallet stones in such a way as to make the inside corners of the pallet stones reach over 3 teeth of the escape wheel, and to make the outside corners of the stones reach over 2 teeth and 3 spaces of the wheel, with a small amount of clearance in each instance, which is called the "drop".

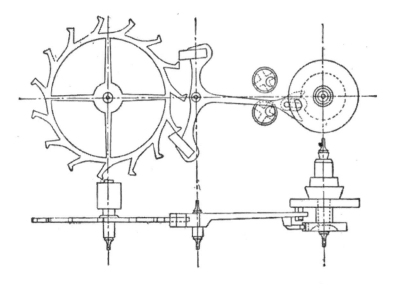

FIG. 76.—LEVER ESCAPEMENT OF THE DOUBLE ROLLER TYPE.

One other important point in relation to the slotting is to direct the slots in the pallet in such a way as to make the locking faces of each pallet stone present to the locking corners of the escape wheel teeth a certain angle of "draw" when the stones are in the position of "lock". Suppose that the escape wheel is being forced in the direction indicated by the arrow, but is prevented from turning in that direction because the locking face of the *R* pallet stone is directly in the way of a tooth. The particular tooth which is resting on the pallet stone is exerting a certain pressure directly towards the pallet arbor. If the locking face of the pallet stone were along the line *B*, which is at a right-angle to that line of pressure, there would be no tendency for the pallet to turn in either direction, but being along the line *C*, which forms an inclined plane in relation to the direction of the pressure, the pressure applied by the escape wheel tooth will tend to pull the pallet stone towards the escape wheel. This action is called the "draft" or "draw". The turning of the pallet is, however, limited by the banking pin, and the object of the draw is to keep the fork against the banking pin all the time that it is not in engagement with the jewel pin. This action of "draw" is similar on the *L* stone; the only differences are, first, that the pressure of the escape wheel tooth is exerted in the direction *away* from the pallet arbor, instead of towards it, and second, that the turning of the pallet, which in this instance is in the opposite direction, is limited by the other banking pin.

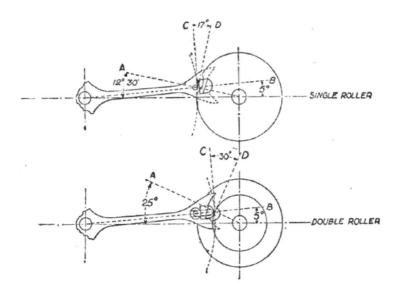

FIGS. 77 & 78.—THE SAFETY ACTION.

A glance at the drawing will make it apparent that the impulse face, which is formed by the surface between the locking and the let-off corners, is at a different angle on the *R* from that on the *L* stone. The impulse angles of the stones in the escapement represented by this drawing are, on the *R* stone, 6° 30⊠, and on the *L* stone 14°. The above refers in each instance to the angle of the impulse face in relation to a right-angle to the locking face, or to the body of the stone. From this condition and from the direction of the pallet stones in relation to the body of the pallet, the factory names "straight" and "crooked" have been given to the *R* and *L* stones, respectively. In books and treatises on the lever escapement the names "receiving" and "discharging" are used, but when, as a matter of fact, both pallet stones perform the functions of receiving and discharging, one as much as the other, these names do not seem to be appropriate. For my part I prefer to use the letters *R* and *L* to distinguish one stone from the other, and these letters may easily be remembered as right and left, this being the order in which the stones appear as we look at the top of the pallet with the fork turned towards us.

The impulse or lift is divided between the escape wheel clubs and the pallet stones; the two together cause the pallet to turn through an angle of 8°30ꞌ.

The lock amounts to from 1° to 1° 30ꞌ, making the total angular motion of the pallet about 10°. This is the condition when the pallet is "banked to drop", that is, when the teeth of the escape wheel will just barely pass by the let-off corners of the pallet stones as the fork comes to rest against the bankings. A certain amount of clearance, or freedom, has to be added to this to allow for oil, etc., so that the bankings have to be turned away from the centre line a small amount to allow for what is called "slide", that is, the pallet stone will slide a visible amount into the escape wheel, after the escape wheel tooth drops on to the same. The amount of slide should, however, be very small, because it causes loss of power, by increasing the resistance to unlocking as, in order to un'ock, the escape wheel actually has to be turned backwards against the power of the mainspring, and the amount of this recoil is in proportion to the lock and slide added together. It is therefore important to notice the action of every tooth of the escape wheel on both pallet stones, to ascertain that each pallet stone has some slide on every tooth, and to allow only a small amount in the place where it appears to be closest.

Roller and Jewel Pin.—One problem in connection with lever escapements, with which every watchmaker has had more or less experience, has come to a final solution in the modern double roller escapement. This is the fastening of the jewel pin. The roller, which holds the jewel pin, is made of bronze, with a hole in it the shape of the pin, but a certain amount smaller than the pin. The jewel pin is made of sapphire, and is made slightly tapering, and is forced into the hole in the roller, thus making it permanently secure. The shape of the jewel pin is round, with one side flattened off to measure three-fifths of the diameter of the pin, and the sharp corners removed. This form of jewel pin is superior in general practice to any other iorm, as it unites strength with the most desirable shape at the points of action. The principle of setting the jewel pin directly in the roller, without cement, is made possible by the double roller escapement, because of the special roller for the safety action. It would not be practical to set jewel pins without cement in steel rollers, as it is in bronze, neither would it be advisable to use bronze rollers for the safety action, because it has been found by experience that tempered steel is better for that purpose. But by separating the two features, it makes a most desirable combination to use a bronze roller for carrying the jewel pin, and a separate steel roller for the safety action.

Matching the Escapement.—The term "matching the escapement" is used to designate the work of bringing the different parts of the escapement into correct relation to each other; in other words, to make the necessary moves in order to obtain the proper lock, draft, drop, slide, fork length, let-off, etc. The best way of learning to do this work is to have a competent instructor who is at hand ready to inspect and to give advice. The difficulties are not so great in *doing* this work, as in correctly determining *what* to do, in order to bring about certain results, and also to know when the escapement is in a proper condition. It is difficult to give in writing a comprehensive idea of how to do this work. I will, however, give a few points which I hope will be useful to the beginner.

The first thing to receive attention is the condition of the pivots on the escape pinion, pallet arbor and balance staff, to see that they are straight, and that they fit properly in their respective holes. It is abolutely necessary that each pivot should have *some* side-shake, but it is also very important to *guard against too much side-shake*, as such an excess causes loss of power and uncertainty in the action of the escapement. A desirable amount of end shake should be from .02 to .05 mm. As soon as these points have been found to be correct we are ready to try the "lock" and the "drop". In describing the pallet action, we made the clear statement that the lock should amount to from 1° to 1° 30⊠. This statement is, of course, of no practical use unless we are equipped with the necessary instruments for measuring this angle. We may, however, use the thickness of the pallet stones for comparison and obtain practically the same results, by making the amount of lock equal to 1/10 to 1/8 the thickness of the stone, from the locking to the let-off corner. This corresponds very closely to the above angular measurements. If the pallet stones are to be moved, in order to change the amount of lock, it is very important to first consider what will be the effect of a certain move, besides the alteration of the lock. The drop, for example, is effected very rapidly by moving the *L* stone. Hence if the drops are equal, we should make the change in the lock by moving the *R* stone. If the lock is too strong, and the drop is largest on the outside, the *L* stone should be moved. If the lock is too strong, and the drop is the largest on the inside, it is necessary to move both stones. Move the *L* stone out a small amount, and move the *R* stone in until the lock is correct. It is also well to recognise that the drop may be modified to a certain extent by moving the pallet stones, close to one or the other side, in the slots; as there is always some room allowed for the shellac which is used for holding the stones. The moving of the pallet stones in or out in the slots will also affect the draft feature of the escapement; this is a point which we should bear in mind whenever we make a change in the position of the pallet stones. The effect of moving the *R* stone out is to increase the draft on both stones, whereas if the *L* stone is moved out and the *R* stone in, it will decrease the draft. In order to ascertain that the escape wheel is correct, the lock and the drop should be tried with every tooth in the wheel on both pallet stones. This should be done with the bankings adjusted close, so as to just permit the teeth to drop. And the best way to try this, is to move the balance slowly with the finger while the pallet action is observed through the peepholes. After completing the adjustment of the pallet action the jewel pin action is next to be considered. The fork should swing an equal distance to each side of the centre line when the pallet is banked to drop. If we find that it moves farther on one side than the other, it will be necessary to bend the fork close to the pallet a sufficient amount to bring it in line. This is called "adjusting the let-off". The test for the let-off is to see that when the pallet is banked to drop, the jewel pin is just as close to the corner of the fork, in passing out, on one side as on the other. This test is correct, provided that the fork is of equal length on both sides of the slot, as it should be. The test for the fork length is that it should allow the jewel pin to pass out on both sides when the pallet is banked to drop. This is the maximum length which is allowed for the fork.

The test for short fork is to move the balance so as to unlock the pallet, then reverse the motion and see that the pallet is carried back safely to lock by the jewel pin. This should be tried on both pallet stones. It is, however, customary to try the shake of the fork when the centre of the jewel pin is opposite the corner of the fork, and not to allow the pallet to unlock from this shake. In order to ensure perfect freedom in the jewel pin action, the jewel pin should be from .01 to .015 mm. smaller than the slot in the fork. The safety action is also adjusted, while the escapement is banked to drop. The guard pin should be made just barely free from the roller when the fork is against the banking, and this should be tried carefully on both sides. If this is done correctly, the roller will have the necessary clearance when the bankings are opened to allow for the slide.

The operation of moving a pallet stone is one that requires a great deal of experience before one is able to do it satisfactorily except by repeated trials. Special tools called "pallet warmers" have been devised for holding the pallet during this operation. In the simplest form this tool consists of a small metal plate, about as large as a 12-size barrel, with a wire handle by which it is held while it is heated. This plate should have one or more holes drilled in it as clearance for the pallet arbor. An improved form of this tool is shown in Fig. 78A. The pallet is placed top-side down against this plate, and the whole of it is warmed over the alcohol flame until the shellac is softened so the stones can be moved.

A good way of applying shellac for the fastening of pallet stones is to warm some stick or button shellac, over a flame and pull it out in long threads of about .5 mm. diameter. Shellac in this form is very convenient to use, as it is only necessary when the pallet is heated to the proper temperature, to touch the end of this thread to it at the place where the shellac is wanted. With a little practice one can learn to deposit just the right amount. After the pallet is cold all shellac on the surface should be cleaned off carefully with a scraper made of brass or nickel.

The Jewel Pin Action.—The fork and jewel pin action involves two distinct functions; the impulse and the unlocking. In order to illustrate and make this statement clear, we will consider the different parts of the escapement in a normal position, a, Fig. 75. The hairspring, controlling the balance, has brought the fork, by means of the jewel pin, to the normal position of rest.

This leaves the pallet in a position where the impulse face of an escape wheel tooth will engage the impulse face of one or the other of the pallet stones, in this instance the *R* stone. Assuming the parts to be in this relation to each other, it is evident that when power is applied to the escape wheel, the escape wheel tooth which is engaging the *R* stone will cause the pallet to turn on its pivots, and this impulse is transmitted to the balance by the fork acting on the jewel pin. The impulse being completed, the escape tooth drops off from the *R* stone, and the second tooth forward comes to lock on the *L* stone, with the fork resting against the banking, Fig. 76. The fork slot is now in such a position that the jewel pin may pass out perfectly free, and this condition is necessary because the impulse which was given to the balance imparted to that member a certain momentum, causing it to continue to turn in that direction until this momentum is overcome by the tension of the hairspring. During this part of the motion, which takes place after the impulse, the jewel pin leaves the fork entirely, but the instant that the momentum in the balance is overcome by the tension in the spring, the balance will start to turn in the opposite direction, the tendency of the spring being to bring the jewel pin to the centre line. Before reaching this point, however, the jewel pin has to perform the very important function of unlocking. At the completion of the impulse we left the fork resting on the banking, with the fork slot in such position that the jewel pin *passed out* perfectly freely, and, figuring on the assistance of the draft and safety action, which will be explained later, we are justified in expecting that the jewel pin shall *pass in* to the fork slot perfectly freely. The instant the jewel pin has entered the slot, and comes in contact with the fork, the work of unlocking begins. And here is to be noticed that for every tick of the watch, the pallet and fork is started from the condition of rest, by a sudden blow of the jewel pin. And not only the pallet is started, but the *whole train has to be started in the reverse direction*, against the power of the mainspring, to unlock the escape wheel in order to receive another impulse. The jewel pin passing out on an excursion, the same as on the other side, returns to unlock, receives a new impulse, and so on, at the rate of 18,000 times per hour. In view of the above it is evident that lightness, as far as it is consistent with strength and wearing quality, is an essential feature in the construction of the several parts. It was once considered necessary to attach a counterweight to the pallet in order to get it in poise, but with the modern light construction of pallet and fork, it has been proven beyond a doubt that the ordinary form of counterpoise was worse than useless, inasmuch as it involved an added mass of metal whose inertia must be overcome at each vibration of the balance.

The Safety Action.—The function of the safety action is to guard the escapement against unlocking from sudden shocks, or outside influences, while the jewel pin is out of engagement with the fork. In the lower grades of watch movements this guard duty is assigned to the edge of the table roller and the guard pin. The passing hollow, a small cut in the edge of the roller, directly outside the jewel pin, allows the guard pin to pass the centre line during the jewel pin action. This form of safety action is called "single roller" and is shown in plan in Figs. 77 and 78. As will be seen from this drawing, the edge of the roller is made straight, or cylindrical, and the guard pin is bent in such a way as to present a curved portion to the edge of the roller. The advantage gained from this construction is that the guard pin can be adjusted forward or back by simply bending it at the base, without its action being in any way affected by a reasonable amount of endwise movement of either the balance staff or the pallet arbor. The double roller escapement, Fig. 76, presents a more desirable form of safety action, for two reasons: first, the intersection of the guard pin with the roller is much greater, making it perfectly safe against catching or wedging; second, any shock, or jar, causing the guard pin to touch the roller, will have less effect on the running of the watch, because the impinging takes place on a smaller diameter. The diagrams, Figs. 77 and 78, illustrate the above statements. The wedge action of the guard pin, when it is brought to the roller, is represented by the lines *C* and *D*, which are at right-angles to the lines *A* and *B*, thus forming tangents to the points of contact. It will be seen that with the single roller this wedge is 17°, whereas in the double roller it is 30°, a very considerable difference in favour of the double roller.

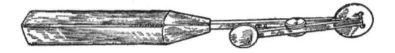

FIG. 78A.—A PALLET WARMER, FOR SETTING THE PALLET STONES.

Directions for Putting the Escapement in Beat.—An escapement is said to be in beat when it requires the same amount of power to start the balance in one direction as in the other. This should be tried with the mainspring only partly wound up, by arresting the motion of the balance with a pointed object held between the heads of two balance screws, and allowing the balance to move slowly, first in one direction and then in the other. If it appears to require more power in order to let off on one side than the other, it is said to be "out of beat", and it should be corrected by turning the hairspring collet a certain amount, on the balance staff, until it takes the same amount of power to let off on one pallet stone as on the other. This is usually done without removing the balance, by reaching in over the top side of the hairspring with a special tool made of small steel wire and flattened at the end so as to enter the slot in the collet. Great care should, however, be exercised in doing this work, so as to avoid bending the hairspring out of true.

THE CYLINDER ESCAPEMENT

The cylinder escapement must not be despised even though it is not to be found in the high-grade watch. This type of escapement is still popular with the cheaper movement where it often gives a considerably better performance than some lever escapements.

As its name implies, the main unit of this escapement consists of a cylinder. The pivoted cylinder, to which is attached the balance wheel, occupies the same place as the balance staff in the lever escapement. The escape wheel, which is of peculiar design, works directly into the cylinder; there is no intermediate connection like the lever or pallets. Brass escape wheels and jewelled cylinders were used by the early makers, but the best results have been obtained by using steel wheels and cylinders.

Unlike the lever, which is a detached escapement, the cylinder is essentially frictional, for the escape wheel teeth rub the edge of the cylinder for a considerable part of the balance vibration. Fig. 82 shows the escape wheel and cylinder, From the sketch it will be seen that the teeth are mounted upon stalks, whilst the plan view reveals the wedge-shaped tooth.

In the action of the escapement, a point of an escape wheel tooth rests on the outside of the cylinder, and as the balance revolves, the tooth forces its way into the cylinder, at the same time giving impulse to the balance. The cylinder, which is, of course, cut away to allow the teeth to enter, now receives the tooth on its inside. There is a little drop as the heel of the tooth leaves the lip of the cylinder.

The tooth remains inside the cylinder until the balance reaches the end of its vibration. When the balance returns it allows the tooth inside the cylinder to escape, and the point of the next tooth to drop on to the outside of the cylinder. In a cylinder escapement, the banking is provided by the balance and the balance bridge. A short pin is fitted into the edge of the balance wheel whilst the banking stud usually consists of a pin attached to the underside of the bridge. Unless the banking stud is close to the rim of the wheel excessive vibration is likely to cause the banking pin to lock itself on the banking stud.

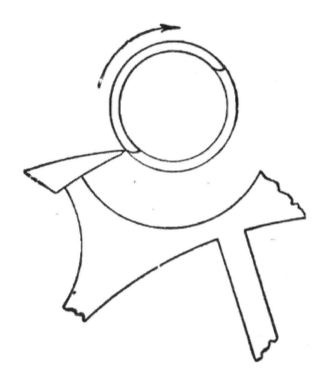

FIG. 79.—THE ESCAPE WHEEL LOCKED BY THE EDGE OF THE CUT-AWAY PORTION OF THE CYLIN-DER.

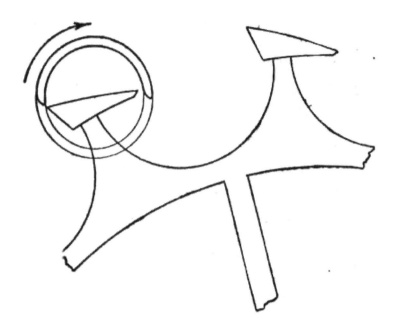

FIG. 80.—AS THE BALANCE SWINGS, THE ESCAPE WHEEL IS UNLOCKED, THE ESCAPÉ WHEEL TOOTH ENTERS THE CYLINDER, AND THE INCLINED FACE OF THE TOOTH GIVES A FRESH IMPULSE TO THE BALANCE.

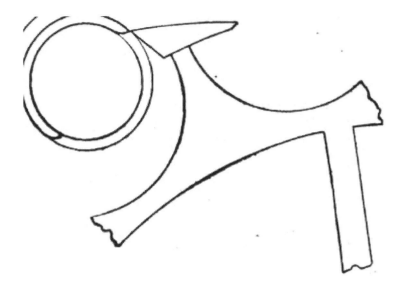

*FIG. 81.—THE TOOTH "ESCAPES" FROM THE CYL-
INDER.*

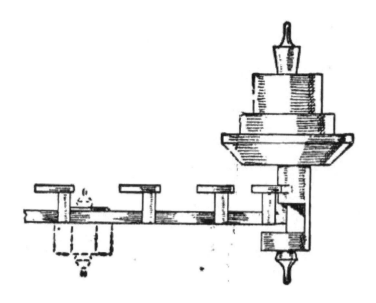

FIG. 82.—THE ESCAPE WHEEL AND CYLINDER.

FIG. 83 (BELOW).—THE CYLINDER AND ITS TWO
END PLUGS ON WHICH THE PIVOTS ARE FORMED.
WHEN A PIVOT BREAKS THE PLUG CAN BE
KNOCKED UP AND A FRESH PIVOT TURNED ON.

High quality escapements are provided with a series of dots to assist in setting the balance in beat. Three dots on the bottom plate and one dot on the balance. When the balance is at rest, its dot should be opposite the middle dot. To test the beat, turn the balance until a tooth drops, note the position of the balance dot in relation to one of the outer dots. Reverse the balance and note its position. If the dots coincide or bear the same relative position to each other, the escapement will be in beat. If the positions are unequal, the hairspring will have to be moved until they become equal.

Should the escape wheel have drop outside the cylinder, but no drop inside the cylinder will be too small, if the conditions are reversed the cylinder will be too large. These faults can only be remedied by fitting a new cylinder. Unequal escape wheel teeth can be equalised by filing the tips of the teeth with a diamond file. The lower cylinder jewel is carried in a movable brass bar called the chariot. In high-class watches this is a separate bar carrying a jewel and an end-piece fixed by a screw or screws. Cheap mass-produced watches have the bottom plate pierced to allow movement to be made, a clumsy but effective method.

By moving the chariot the engagement of the escape wheel with the cylinder can be made deeper or shallower. If the depth is shallow, the balance will trip, and, if the watch continues to go, it will naturally gain considerably. Frequently, the underside of the escape wheel rubs the base of the cylinder opening. To give a clear passage, lay the escape on a hollow brass stake, select a hollow steel punch a little smaller than the rim of the wheel, and stretch the arms of the wheel upwards by delivering one or two light blows with a hammer. Should the lower edge of the cylinder opening rub the upper of the escape wheel, the wheel should be laid upside down on the stake and stretched upwards as previously described.

Printed in Great Britain
by Amazon